John Emery Bucher

An Examination of Some Methods

Employed in Determining the Atomic Weight of Cadnium

John Emery Bucher

An Examination of Some Methods
Employed in Determining the Atomic Weight of Cadnium

ISBN/EAN: 9783337350819

Printed in Europe, USA, Canada, Australia, Japan

Cover: Foto ©berggeist007 / pixelio.de

More available books at **www.hansebooks.com**

An Examination of some
Methods Employed in Determining the Atomic Weight of Cadmium.

A Thesis
Presented to the Board of University
Studies of the Johns Hopkins University
for the Degree of Doctor of Philosophy

by
John E. Bucher.

1894

Contents.

Acknowledgement.

The author wishes to acknowledge his indebtedness for advise and instruction to Professor Morse at whose suggestion and under whose guidance this work has been carried on. He also wishes to express his thanks for instruction to Professor Remsen in Chemistry, Professor Williams in Mineralogy, Dr. Ames in Physics and Mr. Hulburt in Mathematics.

Introduction and Historical Statement.

The atomic weight of cadmium has been investigated by a number of chemists but the results obtained vary between wide limits. The work described in this paper was undertaken with the object of finding the cause of the discrepancy in some of the methods employed. A complete historical statement has been given by Morse and Jones, Amer. Chem. Jour., 14. 261.) and it is only necessary, here, to give a summary for the purpose of reference:

	Ratio.	(A. Wt. Cd.
1818, Stromeyer,	$Cd : CdO$	111.483
	(Schweigg[er's] Jour. 22, 336.)	
1857, Von Hauer,	$CdSO_4 : CdS$	111.735

(Jour. f. Prakt. Chemie 72, 338.)

1859, Dumas, 1st series. $CdCl_2 : Ag$ 112.716

       2d   " $CdCl_2 : Ag$ 112.661

(Ann. Chim. Phys. [3], 55, 158.)

1866, Lenssen, $CdC_2O_4 : CdO$ 112.643

(Jour. f. Prakt. Chem. 79, 251)

1882, Huntington and Cooke, $CdBr_2 : AgBr$ 112.239

          "   " $CdBr_2 : Ag$ 112.245

(Proceedings Amer. Acad. 17, 28)

1890, Partridge, 1st series $CdC_2O_4 : CdO$ 111.811

       2d   " $CdSO_4 : CdS$ 111.727

         "   3d " $CdC_2O_4 : CdS$ 111.611

(Amer. Jour. Sci. [3], 40, 371)

1892, Morse and Jones, 1st method, $Cd : CdO$ 112.616

       2d   " $CdC_2O_4 : CdO$ 112.632

1892, Lorimer and Smith, $CdCl : Cd$ 112.655

(Zeit. f. Anorg. Chem. I. 364)

In this summary as well as in
the rest of this paper the *following*

Atomic weights are used:

Oxygen = 16.00   Chlorine = 35.45

Sulphur = 32.059   Bromine 79.95

Carbon = 12.003   Silver = 107.93

Preparation of Pure Cadmium.

"Cadmium met. puriss. galv. reduc", obtained from Schuchardt, was used for preparing pure cadmium. It was heated to redness in a current of hydrogen which had been purified by washing with both acid and alkaline solutions of potassium permanganate. This treatment converted the metallic powder into a bar which could be distilled in a vacuum. The metal was then distilled

nine times in the same manner that Morse and Burton, Amer. Chem. Jour. 12, 219, had distilled zinc. All distillations were made slowly except the last one, which was made quite rapidly.

Preparation of Nitric Acid.
Whenever pure nitric acid was required, it was purified by distilling against a platinum dish and collecting the distillate in a smaller one of the same metal. The nitric acid used was dilute and free from chlorine.

Purification of Water.

The water used in this work

was purified by distilling twice from an alkaline solution of potassium permanganate, always rejecting the first part of the distillate. Whenever water was needed in the preparation of a pure compound e.g. cadmium oxalate, oxalic acid, cadmium nitrate, etc., it was subjected to the additional process of being distilled against a large platinum dish which was kept cool by placing ice on the inside of it.

Purification of Oxalic Acid.

Commercial oxalic acid was heated with a fifteen percent solution of hydrochloric acid until

all was dissolved. The solution
was then warmed for twenty
four hours. On cooling, crystals
of oxalic acid separated out and
these were washed with a little
cold water to remove the mother liquor.
They were then dissolved in hot
ninety-five percent alcohol and
allowed to crystallize slowly on
cooling. The acid was next
crystallized from ether in which
it is only sparingly soluble.
After this it was boiled with
water until the odor of ethyl
acetate had disappeared. Finally
it was recrystallized three times
from water and dried in the
air at ordinary temperatures.

Preparation of Cadmium Oxalate.

A weighed piece of cadmium was dissolved in nitric acid and the excess of acid evaporated off. The nitrate was then dissolved in a large quantity of water and an equivalent amount of oxalic acid in solution added. The oxalate separated in a few moments as a crystalline precipitate. It was collected on a porcelain filter and washed thoroughly to remove nitric acid and ammonium nitrate. A considerable amount of ammonium nitrate was formed during the solution of the cadmium in nitric acid.

The oxalate was finally dried in
an air-bath for fifty hours, at
150°C.

The Oxalate Method.

Enough cadmium oxalate
for a determination was placed
in weighing tube which had
been tared against a similar
vessel and dried at 150°C. until
the weight remained constant.
It was then poured into
a weighed porcelain crucible.
. The tube and it tare were
now dried again at the same
temperature to constant weight
in order to avoid any error
resulting from moisture being
absorbed by the cadmium oxalate

which adhered to the weighing
glass. The crucibles used in
these determinations were
arranged in the same man-
ner as those employed by
Morse and Jones in their
work on this method. A
small porcelain crucible on
whose edge were placed three
short platinum wire bent in
the shape of the letter U, was
placed in a larger porcelain
crucible. The platinum wires
prevented the lid from sticking
to the crucible after heating
and also allowed the products
of decomposition to escape. The
glaze was removed from the out-
side of the larger crucible with

hydrofluoric acid. To avoid sticking
when heated to a high temperature. A second pair of crucibles arranged in the same
manner was tared against
the first one and in all
cases treated like it. After
the oxalate had been poured
into the weighed crucible, it
was decomposed by placing the
crucible with its contents in
a cylindrical, asbestus-covered
air-bath, and slowly raising
the temperature until the
mass became uniformly brown
in color. In the last five determinations, the temperature was not
allowed to exceed $300°C$ and after
from forty to eighty hours. the

the loss in weight was about ninety percent of the amout calculated for complete decomposition. In the first four the temperature was much higher and the time employed shorter. After the oxalate had been thus treated nitric acid was added and the contents of the crucible dissolved completely. The crucible was then transferred to a bath constructed by placing a larger porcelain crucible in a still larger one of iron and filling the intervening space with sand. It was slowly heated until the nitric had all evaporated and the dry nitrate began to give off red fumes. The cru-

cibles were then removed to a
similar bath containing iron
filings instead of sand. This
bath was heated by means of a
single burner as long as red fumes
were observed, and then for about
five hours with a triple burner.
Finally, the crucibles were trans
fered to a nickel crucible in
the bottom of which a plate of
unglazed porcelain was placed.
The nickel crucible which had
previously been set tightly into
a hole cut in an asbestos board
was then heated over the blast
lamp for two hours. After this
the porcelain crucible and
contents were weighed, and
then reheated for half hour periods

as before until three successive
weighings remained constant.
This usually required from three
to four hours of blasting. In all
determinations, the resulting
product was tested for oxides
of nitrogen with potassium
iodide, starch and hydrochloric
acid, but none were found.
All weighings were reduced
to the vacuum standard on
the assumption of 8.4 for the Sp. Gr.
of brass, 21. for platinum, 3.31 for the
oxalate and 8.15 for cadmium oxide.
The results are:

	Cadmium Oxalate.	Cadmium Oxide	At. Wt. Cd
I	1.97614	1.26414	111.73
II	1.94912	1.24682	111.82
III	1.96186	1.25886	111.77

IV	1.87099	1.19675	111.77
V	1.95941	1.27242	111.79
VI	1.37550	.87994	111.85
VII	1.33313	.85308	111.95
VIII	1.94450	1.24452	112.04
IX	2.01846	1.29211	112.09

A glance at these results shows that there is a variation of .36 of a unit and that the atomic weight in general increases with the number of determinations. In the first four determinations, there may have been loss of cadmium by reduction and subsequent volatilization but in the later determinations this is not probable. It is believed that the greater part of the variation

was due to imperfect dehydration of the oxalate. This and
other sources of error in this
method will be referred to later.
The nickel crucible used gave
a slight sublimate on heating,
even after fifteen hour blasting.
This condensed on the porcelain
crucible as a brownish coating
but, as both the crucible and
its tare were blasted for the
same length of time, it did
not seem to change the difference
of their weights. More than a
dozen nickel crucibles were
tried but none were found not
to give a sublimate. The amount
was so slight that no attempt
was made to determine its nature.

III. The Sulphide Method.

This method is based on the conversion of cadmium oxalate into cadmium sulphide by heating in a current of hydrogen sulphide. The method has been used by Partridge. His result was 111.61 for the atomic weight of cadmium. ~~Earlier, Lovsson obtained~~ 112.14 ~~for it~~.

Preparation of Hydrogen Sulphide

In the present work was always prepared from potassium hydrosulphide which was made from barium sulphide (commercial). Barium sulphide was treated with dilute hydrochloric acid and the resulting hydrogen sul

phide washed thoroughly with
a solution of potassium hydro
sulphide and then with pure
water. It was then passed
into a strong solution of potas
sium hydroxide until the later
was saturated. When it was
required, it was set free from
this solution by adding dilute sulphuric acid and again
washing the resulting gas with
a strong solution of potassium
hydrosulphide.

Preparation of Nitrogen.

Whenever a current of nitrogen
was required, it was prepared
by passing air over a layer of hot
copper gauze in a combustion.

tube. A short layer of copper oxide was first introduced, then the copper gauze and finally another layer of copper oxide. The air was dried with caustic potash before entering the tube and the nitrogen obtained was also passed through a long tube filled with lumps of this substance before being used.

Mode of Procedure.

A number of weighing tubes 140 millimetres long and 13 millimetres internal diameter were made especially for this work. They were always used in pairs one being kept as a counterpoise.

A porcelain boat of such dimensions as just to slide into the tube was placed in each one. For a determination, a tube and its boat were taken with another tube and boat, glass against glass and porcelain against porcelain until the difference in weight was less than two tenths of a milligramme. Both boats were heated in a current of hydrogen sulphide to incipient redness for about one hour. The current of hydrogen sulphide was then replaced by one of nitrogen, in which the boats were cooled but while still warm they were transferred to their weighing tubes and allow-

ed to cool in a desicator containing caustic potash and weighed. Before weighing the stoppers of the weighing tubes were loosened for a moment in order to equalize the internal and external pressure. This treatment was usually repeated two or three times and the difference in weight remained perfectly constant. A portion of cadmium oxalate sufficient for a determination was placed in the weighed boat and dried at $150°C$. The oxalate had been prepared exactly like that used in the oxalate method which has already been described. The gas pressure in the laboratory varied

very much while this method
was under investigation and
great difficulty, was experienced,
in maintaining a constant
temperature although a thermo-
regulator was used. Sometimes
a specimen of oxalate which
was supposed to be dry would
loose several tenths of a milligramme when the thermometer
would only have gone up to
$110°C.$ or $115°C$ for an hour, by
accident. Under these conditions
the drying was so uncertain
that only four determinations
were completed although many
were started. The boat containing the oxalate which had
been dried and weighed was

placed on supports of unglazed
porcelain in a combustion
tube and current of dry hydrogen
sulphide passed over it.
As soon as the air was
expelled, the tube which was
in a combustion furnace,
was slowly heated until all
the oxalate seemed to be
decomposed and then raised
to dull redness. After this
temperature had been maintained for about an hour,
the sulphide was allowed to
cool to a temperature of about
200°C. and the current of hydrogen sulphide replaced by dry
nitrogen, using a three way-
stopcock. When nearly cold the

boat was slipped into its weighing-
-tube and weighed, the same
precautions being used as
when weighing the empty boat.
At this stage the sulphide was
always from one to two milligrammes lighter than at the
end of the determination.
It was reheated for periods
of one hour until the weight
remained constant. This gener-
ally required from three to five
hours. All weighings were
reduced to the vacuum stand
ard on the basis of 4.5 for the
Sp.Gr. of cadmium sulphide,
3.31 for the Sp. Gr. of cadmium oxalate,
8.4 for the sp. Gr. of brass weights
and 21 for the Sp.Gr of platinum weights

The results are as follows:

	Cd C$_2$O$_4$	CdS	At. Wt. Cd
I.	2.56319	1.84716	112.23
II.	2.18364	1.57341	112.17
III.	2.11643	1.52462	112.05
IV	3.13105	2.25582	112.12

The first three determinations were made exactly as above described, the heating in hydrogen sulphide being done in a Bohemian glass combustion tube. The hydrogen sulphide was dried with calcium chloride.

The fourth determination was made under somewhat different conditions. The boat containing the weighed oxalate

was placed in a combustion tube which passed through an asbestos covered air-bath. The air was displaced by a current of dry hydrogen sulphide and the bath slowly heated. When the temperature had risen to 210°C. it was maintained there for three hours, and then raised to 250°C for three hours. The sulphide then weighed 2.27 grammes, being 14 milligrammes heavier than when the determination was finished. It was replaced in the tube and reheated in a current of hydrogen sulphide at a temperature of 300°C. for four hours. It was then transferred to a porcelain tube and heated to

redness for one hour. It then weighed 2.25437 grammes, being 1.45 milligrammes lighter than at the end of the determination. The weight did not become constant until it had been heated six hours more to redness in a current of hydrogen sulphide. When this oxalate was slowly heated, in H_2S a small amount of oxalic acid sublimed to the colder part of the tube, but, in the other cases where the heating was more rapid, only carbon monoxide, carbon dioxide and water were observed.

Discussion of the Method.

When hydrogen sulphide is passed

through a red-hot tube, sulphur
is deposited on the colder parts
because at this temperature hydrogen sulphide dissociates and the
element do not recombine on
cooling. In this work a faint
sublimate was noticed before
coming to the zone of sulphur deposit. On exposure to air, it
deliquesced in a few minutes
forming small yellow drops
which had a saline taste, and
gave tests for potassium and
sulphur. The sublimate had a
yellow color and was evidently
formed by the action of sulphur
on glass. It seemed to do no
harm, but in the fourth determin
ation an effort was made to avoid

it by using a porcelain tube instead of a glass combustion tube
for heating to redness in a current of hydrogen sulphide.
The fact that sulphide of cadmium
was always too light after the
first hour's heating in hydrogen
sulphide proves that it must
have contained some oxide of
cadmium even after this heating.
Oxide of cadmium is very readily
absorbed by the glaze on porcelain,
and some error must have been
introduced in this way because
it would not be converted into
sulphide after forming a silicate.
The effect of this would be to
give a low result for the atomic
weight of cadmium. To get some

idea of the magnitude of this
error, the sulphide was poured out
of the boats used in the first and
second determinations. They
were then warmed with nitric
acid for a few moments, washed
in water, and heated over the
blast-lamp for a few minutes.
The boats used as tares were
treated in exactly the same
manner. On weighing, the boats,
in which the oxalate in determinations I and II had been decomposed, were found to be 1.12 milligrammes and .82 milligrammes
heavier respectively than at
the beginning of the determinations
This would only introduce an
error of .03 % unit in the

atomic weight on account of the
small difference in weight between
these amounts of oxide and
equivalent amounts of sulphide.
The boats were warmed, as above
mentioned, with nitric acid
to remove any adhering sulphide.
This might have decomposed
some cadmium silicate at the
same time, and the error due
to cadmium oxide thus be found
smaller than it really is.
The following experiment was
made in the hope of avoiding
the formation of cadmium silicate:
The glaze was removed from
the inside of a porcelain boat
by hydrofluoric acid followed by a
thorough scouring with sand and

and water. The boat was then heated in the flame of a blast lamp for several minutes, tared against another boat which was not treated with hydrofluoric acid. Both were heated to redness in a current of hydrogen sulphide for an hour, cooled, weighed, and then heated in hydrogen sulphide for another hour, and weighed again. The boat gained 1.7 milligrammes during this second heating, showing that a boat whose glaze has been removed by hydrofluoric acid could not be used in this method. Throughout this work, great care was taken to exclude

the oxygen of the air from the cadmium sulphide, while hot. The current of hydrogen sulphide in which the cadmium sulphide is heated must not be too slow, otherwise the sulphur in the dissociated gas will diffuse to the colder parts of the tube and condense; the residual gas becoming very rich in hydrogen. The hydrogen will then reduce some of the sulphide to metal, causing loss by volatilization. One determination was lost in this way, over two milligrammes of the sulphide being sublimed out, and it could easily be detected, on the side of

the tube. It is believed that
the cause of the variations
in the four determinations
made by this method, is due
to imperfect dehydration of
the oxalate. It did not seem
advisable to continue this part
of the work any farther; therefore
the chloride method was taken
up.

The Chloride Method.

Huntington had determined
the ratios of $CdBr_2$ to $AgBr$ and
also $CdBr_2$ to Ag very carefully, attain
ing the result 112.24 for the atomic
weight of cadmium. Morse and
Jones had obtained 112.07 for this

constant by the oxide method.
The object of the work about
to be described was to find
the cause of this discrepency
if possible. It was thought advisable however to make some
determinations of the ratio
of $CdCl_2$ to $AgCl$ before beginning
the bromide method.

Dumas, in 1859, used cad-
mium chloride to determine
the atomic weight of the
metal. He did not establish
its ratio to silver chloride but
to silver by titration. He prepared
cadmium chloride by dissolving
the metal in hydrochloric acid
and melting the resulting
product in a platinum capsule

for five or six hours. He made two
series of three determinations
The chloride used in the first series
was yellow in places and not
completely soluble. The result was
112.476. The second series was
made with chloride which was
perfectly white and soluble and
gave 112.007 for the atomic weight
of cadmium. It is evidently more
reliable than the first series and
Dumas himself concluded that
the atomic weight is very near
112.0.

Preparation of Cadmium Chloride

Four different specimens
of cadmium chloride were used
in this work and from these

specimens portions were taken for analysis. These portions were treated differently in different analyses. therefore it will be necessary to give a brief description of them and mention the number of the determinations, in which each one was used. Chloride of cadmium was prepared in the following manner. A solution of pure hydrochloric acid was prepared by passing a current of hydrochloric acid gas into pure water which was contained in a porcelain crucible until no more was absorbed. The water used had been purified by distilling against a platinum dish and the

hydrochloric acid gas was obtained by heating ordinary concentrated chemically pure hydrochloric acid in a distilling-bulb whose neck had been closed by fusion in order to avoid the use of a cork or rubber stopper. Hydrochloric acid thus prepared will leave no residue on evaporation when air is excluded (Stas, Aronsteins German translation, p. 111). A piece of platinum foil freed from iron by heating in the vapors of ammonium chloride as recommended by Stas, Aronsteins translation, p. 112) was introduced and a piece of cadmium laid on it. Solution begins at once, the hydrogen being liberated on

the platinum foil. During the
later part of the process, heat
was applied. After all of the
metal had dissolved the solution
was evaporated, the platinum
foil having previously been
removed. The crystals of cadmium chloride which separated
were not dried but allowed to
remain slightly moist with
hydrochloric acid. If no platinum foil is used, the solution
of the pure metal becomes ex
ceedingly difficult, unless a very
large excess of acid is used.
No objection can be raised to the
use of platinum foil for in making
fifty grammes of cadmium
chloride it lost less than a tenth

of a milligramme and even
this could probably have been
avoided by using a somewhat
larger amount of hydrochloric
acid. The foil was always kept
submerged in the acid liquid.
The moist crystals of cadmium
chloride were transferred to a
combustion tube passing through
an asbestus covered air-bath,
and dried in a current of hydrochloric acid gas for several hours
at 300°C. The hydrochloric acid gas
had been passed through a long
calcium chloride tube to dry it,
although calcium chloride
probably does not do this very
thoroughly. The hydrochloric
acid gas was then replaced by

a current of nitrogen prepared as
has already been described under
the sulphide method. After the
current of nitrogen had been
passing for about half an hour,
the tube was allowed to cool,
and the chloride transferred
to another combustion tube, one
end of which had been sealed
in the flame of a blast-lamp.
The other end was drawn out
and attached to a Sprengel
mercury pump. After exhausting,
the chloride was sublimed in
the vacuum. This takes place
at a moderate temperature and
the sublimate has a beautiful
crystalline structure and is
perfectly white.

The crystalline mass exposes so much surface that water is taken up very rapidly when exposed to the air. This action is so rapid that the crystals cannot be transferred to a weighing glass without introducing an appreciable error. The whole sample was accordingly transferred to a stoppered glass bottle which was kept under a bell-jar with sticks of caustic potash. Three samples were prepared in this manner, the first being used in determination one, the second in determinations two to seven inclusive, and the third in determinations eight to nineteen inclusive.

The samples used in determinations twenty and twentyfone
were prepared in the following
manner: About three grammes
of cadmium were placed in a
combustion tube in which three
bridges (as in the distillation
of pure cadmium) had been
made. A section may be represent-
ed thus

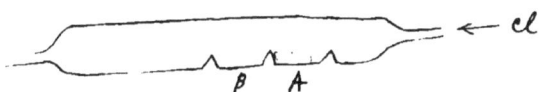

The metal was placed in
cavity A and a stream of
chlorine passed through the
tube. The chlorine was prepared
from potassium bichromate and
hydrochloric acid and dried

by passing it through a long tube
containing calcium chloride.
When the air had been displaced,
the cadmium was heated. It
fused and began to burn to the
chloride which partly flowed
over the bridge into cavity B
and partly distilled over into
this cavity. When the reaction
had ended, the current of
chlorine was replaced by one
of dry nitrogen, and the tube
was allowed to cool, and the
chloride taken for analysis XX.
The specimen used in analysis
twenty-one was prepared in ex
actly the same way, only the
chlorine used was obtained from
manganese dioxide. sodium chloride

and sulphuric acid, and was dried with phosphorous pent-oxide instead of calcium chloride.

The special treatment of the portions taken for analysis was as follows: Those taken for determinations I. II and from XI to XIX inclusive were placed in a platinum boat and put into combustion tube. A current of hydrochloric acid gas obtained by heating the aqueous acid was passed through the tube. The gas had been dried by calcium chloride. When the air was displaced, the choride was heated somewhat higher than its fusing-point

i. e. to incipient redness, and
maintained there for a length
of time varying from a few
minutes to more than an
hour. The hydrochloric acid
was then displaced by a current
of nitrogen, and the chloride
allowed to cool. The boat with the
chloride, while still slightly
warm, was placed in a weighing-tube, cooled in a dessicator
containing caustic potash and
weighed. The chloride thus
prepared is transparent and
presents only a small surface
to the air. It takes water up so
slowly that no error is introduced from this source. This
was tested in one case by

allowing a boat containing some chloride thus prepared to stand in the air for a certain length of time, and noting the increase in weight. It was quite slow. In several cases specimens of chloride were tested for hydrochloric acid using tropaeolin as an indicator. It was always found neutral. The portions used for determinations III and from VI to X inclusive were prepared in exactly the same manner as the preceeding ones except that the hydrochloric acid gas in which they were fused was not dried but used just as it came from

the aqueous acid. In some
cases the platinum boat in
which the chloride was fused
was weighed before and after
the fusion. The weight remained unchanged.

For determinations IV and V,
about six grammes of of cadmium
chloride were placed in a
platinum boat and more than
two-thirds of it distilled out
in a current of hydrochloric
acid gas which had not
been dried. Part of the distil
late was collected after cooling
in nitrogen and used in deter
mination IV while the residue
remaining in the boat was
used for determination V.

The method of preparing the chloride used in determinations XX and XXI has already been described.

The Filters.

Thinking that a Gooch crucible with a platinum sponge on the bottom in place of asbestus would be desirable for this work one was according ly made and answered the purpose very satisfactorily. All determinations were made by using such filters. (.E. Munroe (Chem. News, vol 58. p. 101) has described the preparation of these filters.

A platinum Gooch crucible was placed on a filter paper and some ammonium platinic chloride which had been thoroughly washed introduced by suspending it in alcohol and then pouring this into it. The precipitate settles to the bottom forming a uniform layer, and the alcohol drains off through the filter paper. The crucible was then dried slowly in an air-bath. After this it was transferred to a porcelain crucible and slowly heated until decomposition was complete. In this manner a layer of platinum felt is obtained which acts as a very efficient

filter. Another layer of double chloride was then decomposed as before so that if there were any imperfections in the first layer they would be covered by the second layer. The surface was smoothed down by means of a glass rod. To prepare a good filter the drying and subsequent heating should be very slow. The heating must not be at too x high a temperature, otherwise the felt becomes very compact and is useless for filtering purposes. Pressure produces the same effect. The filters were + always treated with strong nitric acid, washed and

reheated before being used, but in no case was chlorine detected in the nitric acid after the washing, nor any loss in weight of the crucible. An objection to the use of these crucibles for the purpose named was found in the course of this work, but it will be discussed later. The crucibles were always set in a large weighing-glass, and another weighing-glass containing an equal amount of platinum foil used as a tare, in weighing. This precaution was perhaps unnecessary, but at least it did no harm.

Analytical Process.

The weighed cadmium chloride was dissolved by placing the boat containing it in an Erlenmeyer flask containing water. The boat was then washed, dried and replaced in its weighing-tube. On weighing again, the loss in weight is equal to the weight of cadmium chloride taken. All samples gave a perfectly clear solution except those used for determinations XX and XXI. A drop of nitric acid (1:3) was added to each solution except in determination XIV where three cubic centi-

metres were added and in
XVI where ten cubic centimetres
were added. A solution of
silver nitrate was then
added to precipitate the chlorine. This as well as the subsequent washing was done in
a dark-room illuminated by
a single gas light whose
rays had to pass through
a strong solution of neutral
potassium chromate. The
precipitate was contracted
by warming on the water-
bath. It was then collected in
the prepared Gooch crucibles,
and washed. Before filtering,
the flask containing the precip
itate and mother-liquor was

allowed to cool. Silver chloride
is soluble in water to a
considerable extent but is
reprecipitated by adding an
excess of either silver nitrate
or hydrochloric acid. Stas (Zm.
de Chim. et Phys [4], 25, 22; [5], 3, 146; [5], 3, 251.).
investigated this very thorough
by. Cooke also did some work
on it and used a dilute solution
of silver nitrate to wash the
chloride thus preventing
solution (Proc. Amer. Acad. 17, 1.).
In the above work therefore
a solution containing 0.10 grammes
of silver nitrate per litre was
first used, followed by one
only one-tenth as strong, and
finally pure water was used.

Only two or three washings could be made with water as the chloride went into solution after this owing to the removal of the silver nitrate. The last silver nitrate solution used is so weak that any error introduced by not washing it out completely is insignificant. After washing, the silver chloride was dried at a temperatures varying from 150°C to 300°C to constant weight. A glass air-bath was used, in order to prevent products from the burning gas from coming in contact with the chloride. It was then weighed.

The quantity of silver nitrate used in the determinations was varied very much. The excess over what was required to precipitate the chlorine is given in the table of results in those cases in which it is known. The quantity of water used in each determination is also given where it is known. It is given in the number of cubic centimetres used per gramme of cadmium chloride and does not include wash-water. All weighings are reduced to the vacuum standard on the basis that Sp. Gr's of $CdCl_2 = 3.94$ and $AgCl = 5.5$ The results are:

no.	$CdCl_2$	$AgCl$	H_2O per Grm	Excess AgCl	At. Wt.	melted in
I	3.09183	4.83856			112.339	Dry HCl
II	2.26100	3.53854			112.329	" "
III	1.35729	2.12431			112.320	Moist HCl
IV	2.05582	3.21727			112.339	" "
V	1.89774	2.97041			112.306	" "
VI	3.50367	5.48473		8.90	112.283	"
VII	2.70292	4.23087	200	1.79	112.301	" "
VIII	4.24276	6.63 98	300	8.10	112.387	" "
IX	3.40200	5.32314	300	18.95	112.368	" "
X	4.60659	7.20386	300	25.62	112.472	" "
XI	2.40832		- - -		112.434	Dry HCl
XII	2.19114	3.42724	- - -		112.433	" "
XIII	2.84628	4.45477	300	4.45	112.319	" "
XIV	2.56748	4.01651	300 +3cc HNO₃	.10	112.399	" "
XV	2.31003	3.61370	300	.10	112.406	" "
XVI	1.25008	1.95652	300 +10cc.HNO₃	4.66	112.319	" "
XVII	1.96015	3.06541	300	3.22	112.466	" "
XVIII	2.29787	3.59391	300	4.27	112.448	" "

Discussion of the Results.

In the first five determinations, the analytical oper-
ations were conducted as
nearly as possible alike, but
the preparation of the portions
of cadmium chloride taken
for analysis was varied very
much as will be seen by
referring back to this part of
this paper. The results do
not vary more than ±0.015 from
their average. This is very

strong evidence of the purity of
the chloride used for, if it contained any impurity, we would
have expected to vary the amount
in the different portions.
After this, attention was
paid especially to the analytical process for it was thought
that there probably was some
serious error in the method.
the result being ~~higher~~ than any
that had previously been obtained. if we exclude Dumas'
first series which he himself
~~did~~ not accept. The conditions
were varied in many ways—
to see how much the result
could be influenced. but
under no conditions were

results as low as Huntington's average (112.24) obtained. A numbers of errors were found in the method during the work, but they seem to neutralize each other to a great extent. The more important ones will now be given. Nearly every filtrate including the corresponding wash water was examined for chlorine after the silver and cadmium had been precipitated by hydrogen sulphide. The excess of hydrogen sulphide was expelled by boiling, after the addition of some nitric acid. In two cases an inverted condenser was used. On adding silver

nitrate a precipitate was always obtained, showing the presence of chlorine. Care was always taken to filter off sulphur formed by the oxidation of hydrogen sulphide, before adding the silver nitrate. The precipitate was never very heavy, and was not estimated quantitatively. It is evident that cadmium nitrate exerts a solvent action on silver chloride. In some cases a very large excess of silver nitrate was added, but it did not change the results markedly. Silver nitrate itself dissolved silver chloride to some extent. The increase in insolubility, if any, on adding

an excess of silver nitrate is
probably counterbalanced by
the increased error due to occlusion of nitrates in the
silver chloride. Stas (Aronstein
Trans. p. 156   ) says it is impos-
sible to contract silver chloride +
or bromide in a solution
containing salts without
there being occlusion and
that the precipitate can only
be freed from them by dividing
up the contracted mass by shak-
ing with pure water. This
was not done here owing to
the solubility of silver chloride
in pure water, and the comp
lications introduced in the
analytical part. The occlusion

of nitrates by the silver chloride
would lower the atomic
weight found. The silver
chloride obtained always
darkened on heating and
contained cadmium, as was
shown in the following manner: The lump of silver
chloride was attached to
the negative pole of a cell
and electrolyzed in a bath
containing dilute sulphuric
acid. The resulting metal
was then dissolved in nitric
acid and the silver precipitated
by adding hydrochloric acid.
The filtrate was evaporated to
expell the nitric acid, and the
residue taken up with water

and tested for cadmium with hydrogen sulphide. An appreciable quantity was always found. This method of examination does not show the occluded silver nitrate. Another error which tends to lower the atomic weight found is due to the platinum crucibles used for filtering. If a silver nitrate solution is filtered through a crucible there will be an increase in weight due to silver being deposited. This takes place in acidified solutions as well as in neutral ones. Washing with ammonia does not remove the deposit, but

strong nitric acid does, the
washings giving a test for
silver. Whether the depositing
of silver is due to the action
of spongy platinum in
contact with the compact
metal of the crucible or
to some impurity in the
platinum sponge was not
determined, but the former
seems by far the most probable. The increase in weight
during the time required
for filtering a determination
must have been quite small
however. The samples of
cadmium chloride employed
for determinations XX and
XXI were prepare by burning

cadmium in a current of
chlorine. The glass tube used
was attached somewhat and
the solution of the chloride
was very slightly turbid in
each case. The turbidity was so
slight however, that no very
serious error could have resulted
from it, particularly as it was
probably partly counterbalanced
by the formation of some potassium
chloride. For more accurate
work, it should have been
made and redistilled in
a porcelain tube. These two
samples were tested for
free chlorine with potassium
iodide and starch paste,
but none was found. Some

of the specimens of chloride
prepared by fusion in a current
of hydrochloric acid were
found to be neutral. using
tropaeolin as an indicator.

As nearly as can be judged,
the above errors would probably counterbalance each
other to a great extent, and
thus give a fairly close
approximation to the atomic weight of cadmium when
the average of all the
determinations is taken.
The value 112.3^{83} thus obtained
can only be regarded as
tentative.

The Bromide Method.

Huntington (Proc. Amer. Acad.
11. 28) working under the direction
of J. P. Cooke, determined the ratio
of cadmium bromide to silver
bromide and using the total
quantities for the calculation
the result for the atomic weight
of cadmium is 112.237. He
also determined the ratio of
cadmium bromide to silver
obtaining 112.245 for the atomic
weight of cadmium.
The work which will now be
described was carried out very
much like the work described
under the chloride method. The
ratio of cadmium bromide to silver

bromide was investigated.

Preparation of Cadmium Bromide.
and Hydrobromic Acid.

A large quantity of hydrobromic acid was prepared
according to the method described by Dr. Edward R. Squibb
(Trans. of Med. Soc. of the State of N. Y.).
One part of water was added
to seven parts of strong sulphuric
acid (Sp. Gr. = 1.53) and the mix-
ture cooled. Then six parts of
potassium bromide were dissolved in six parts of hot
water and the diluted sulphuric acid added to this hot solution. It was set aside until

cold in order to allow the sulphate of potassium to crystallize
out. The crystals were drained
on a filter-plate and quickly
washed with two parts of water.
The mother-liquor and washing
were then distilled until no
more acid was obtained
on further heating. The
acid thus obtained was
distilled three times from potassium bromide, twice from
cadmium bromide formed
by adding a piece of pure
cadmium to it, and twice
without the addition of anything. It was tested and found
to be free from sulphuric acid.
Cadmium bromide was prepared

from it in exactly the same
way that the cadmium
chloride used in the chloride method was prepared
from pure metal and hydrochloric acid. While the
crystalline mass of cadmium
bromide was still moist, it
was transferred to a combustion
tube and dried at 300°C for
several hours in a current
of nitrogen. It was then
sublimed in a vacuum as
the chloride had been. This
specimen served for the first
three determinations. About
nine grammes of it was placed
in a platinum boat in a com
bustion tube, and part of it dis

tilled in a current of nitrogen
The distillate, a portion of
which had been tested with
tropaeolin and found neutral.
was used for determination
I. The residue in the boat
was used for determination II.
Another portion of the main
sample was resublimed in
a vacuum and used in
determination no three. Cadmium &
bromide is not hygroscopic or
at least only slightly therefore
the sublimed cadmium
bromide can be transferred to
a weighing-glass without taking up water. This cannot
be done in case of the
chloride .

It is probable that the hydrobromic acid as above prepared
was perfectly free from hydrochloric acid. Chlorine in cadmium bromide would cause
the atomic weight to be found
lower than it really is. It was
thought desirable, however,
to prepare a acid which would
certainly be free from chlorine.
The method described by Stas
(Aronsteins German translation, p. 134)
was employed, with the addi
tional precaution that the
above purified acid was used
to start with, and all reagents
employed had been 1) specially
prepared so as to be free from
chlorine. Pure silver was pre-

pared according to Stas'
description (see Aronsteins translation
page 34 also page 104) by the action
of ammonium sulphite on
an ammoniacal solution of
silver nitrate and copper
sulphate. The silver was
dissolved in nitric acid
free from chlorine, and then
slowly added to a dilute
solution of the above described hydrobromic acid, and
the precipitated silver bromide
thoroughly washed. It was
then digested for a long while
in a strong solution of potassium bromide, first in the cold
then by heating. The potassium
bromide had been made thus:

Twice recrystallized potassium hydrogen tartrate was heated in a platinum dish in a muffle furnace until it was converted into carbonate, and the excess of carbon burned off. It was then dissolved in water, filtered and neutralized with some of the hydrobromic acid already described. The carbonate had been tested for both sulphuric acid and chlorine with negative results. After the silver bromide had been digested with the potassium bromide it was washed very thoroughly, suspended in water, and a current of hydrogen sulphide passed into it. This converts it into sulphide

hydrobromic acid being liberated.
The acid was drained off on a
porcelain plate, and then
distilled a number of times. It
was finally tested and found
to be perfectly free from sulphates,
and also did not contain free
bromine. Having started with
an acid which was probably pure
and subjected it to these oper-
ations with reagents free from
chlorine, there can be no doubt
as to the purity of the resulting
acid. The hydrogen sulphide
used was prepared from potassium hydrosulphide as in the
sulphide method, and washed
first with a solution of the
hydrosulphide, then very thorough-

ly with pure water. From the
hydrobromic acid obtained a
specimen of cadmium bromide
was prepared as before. and
sublimed twice in a vacuum.
This specimen was used for
determinations IV and V.

Method of Analysis.

The first three determinations
were made exactly like those
in the chloride method. The
last two were also made in
the same manner only the
washing of the precipitate was
varied. After the silver bromide had been contracted
by warming on a water-bath

it was washed by decantation and then agitated violently with cold water, to remove occluded nitrates, but it was then so finely divided that it could not be filtered. The artifice used by Stas to contract it a second time was to pass a current of steam into the milky liquid. This was tried here, but for some reason or other did not work very well, and considerable difficulty was had in filtering it. The results of the five determinations are tabulated below. All weighings are reduced to the vacuum standard on the basis of Sp.Gr. of ClBr - 4.6 and Sp.Gr. AgBr = 6.42.

No	CdBr₂	AgBr	H₂O	Eq. AgNO₃	At.Wt.	Remarks
I	4.39941	6.07204	---	---	112.73	Distillate }
II	3.18030	4.38831	---		112.42	Residue }
III	3.60336	4.97130			112.45	Resublimed.
IV	4.04240	5.58062			112.29	
V	3.60500	4.97319	---	---	112.35	

Average 112.394

Discussion of the Results.

The first three specimens were prepared under widely differenent conditions yet the results agree quite closely. The last two were prepared from the repurified hydro-bromic acid. If chlorine had been removed during the second purification we would expect a higher result

but the results are lower. There
seems to be hardly any doubt
that this is due to analytical
errors rather than a change
in the composition of the
bromide. Whether this be true
or not, the five determinations
all fall within the limits
obtained by the chloride
method and confirms it
as fully as can be expected.
The errors of the method are
the same as those of the
bromide method, only they are
probably less in most cases.
One filtrate was examined for bromine,
but none was found, showing the
method to be more perfect in
this respect.

VI. Syntheses of Cadmium Sulphide.

It was next thought of examining the method based on the conversion of cadmium sulphate into cadmium sulphide, which has been used by von Hauer whose result is 111.94 for the atomic weight of cadmium, and more recently by Partridge who obtained a much lower result, namely 111.73. They dried cadmium sulphate in porcelain boats, and then reduced it to sulphide by heating in a current of hydrogen sulphide. The reduction begins in the cold and is probably complete or at least nearly complete before the temperature is sufficiently high

for cadmium sulphate to decompose into cadmium oxide, for the sulphate is very stable with respect to heat. This being the case, probably no error results from the formation of a silicate of cadmium in this method. The main difficulty in this method would be to prove that the cadmium sulphate used is free from water. neither Von Hauer nor Partridge have done this because drying a substance to a constant weight is not sufficient evidence of its anhydrous character, especially if the drying is done at a constant temperature. This has been shown very clearly in the case

of copper sulphate, by Richards. (Proc.
Amer. Acad. Sci. 26. 263.   )
It was therefore decided to
attempt the synthesis of cadmium sulphate, hoping to be
able to fix a minimum value
for the atomic weight of cadmium.

A piece of hard glass tube
was closed at one end by fusion
and the other end drawn out
into a small tube which was
then bent twice at right angles.
The large part was cut off near
the beginning of the smaller
tube, and the edges rounded by
fusion. It was filled with dilute
sulphuric acid and heated for
some time to remove soluble

matter from the glass. After re
moving this acid, a weighed
piece of cadmium was introduced and an excess of dilute
sulphuric acid (1:3) added. The
tube contained a small piece
of platinum to aid the solution
of the cadmium. During the
process of solution, the two parts
of the glass tube were held togeth-
er by a rubber band, and the
outlet of the smaller tube dipped
under pure water contained in
a small tube closed at one end.
        A section of the
arrangement is shown
in figure 2. Solution
was aided by the
application of heat.

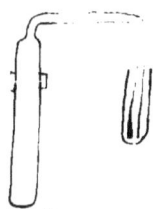

Fig.1

These precautions in dissolving
the metal were taken to prevent
loss by spraying. After the metal
had been dissolved, the solution
and the water through which
the hydrogen had escaped were transferred to a porcelain crucible.
An equal amount of sulphuric
acid was then added to the
tare and both were heated until
fumes of sulphuric acid ceased
to come off. The crucible containing the dry sulphate was was
next placed on a porcelain plate
in a nickel crucible set in a
whole in an asbestus board.
This was placed over the flame
of a Bunsen burner, so that the
bottom of the nickel crucible was

barely at a red heat. The
temperature on the inside of
this bath was considerably lower.
After the weight had become
nearly constant, the sulphate was
tested for sulphuric acid by
means of standard alkali using tropaeolin as an indicator.
It was found acid, but so slightly
that no attempt was made to
estimate it. Result = 112.35 is preliminary.

Another synthesis was
made as follows: A platinum
crucible, lid, and perforated cone
were placed in a large weighing-
-glass and tared with a similar
weighing-glass, containing a platinum crucible, platinum foil being added until the weights were

equal. After these had been accurately weighed, a weighed piece of cadmium was added to the one containing the cone. The cone was inverted over the piece of metal on the bottom of the platinum crucible. A considerable excess of dilute (1:3) sulphuric acid was then added, the lid whose edge was bent down placed on the crucible, and the weighing-glass stoppered loosely. This was placed in an airbath, and gently warmed during the later part of the process of solution. There is no difficulty in getting complete solution if a sufficient excess of acid.

is used. A vertical section of the crucible and weighing glass is shown in figure 3.

This arrangement avoids loss from spraying, and the necessity of transferring the solution from a tube to a crucible as in the first experiment.

An equal quantity of sulphuric acid was added to the crucible used as a tare and evaporated. After the metal had been dissolved the platinum cone was tilted to one side, and the excess of acid evaporated off. It was then heated in a glass air-bath for a long time at a temperature

which was probably about 400°C. After the weight had become constant the amount of free sulphuric acid was estimated by titration with a standard alkali using tropacolin as an indicator. 1.25 milligrammes were found and this weight was subtracted from that found at the balance. Weighing were reduced to the vacuum standard, assuming the Sp. Grs. of cadmium and anhydrous cadmium sulphate to be 8.54 and 3.0 respectively. The results were as follows:

	Cd	CdSO₄	At. Wt.
I.	(112.35 can only be regarded as a preliminary experiment)		
II.	1.15781	2.14776	112.35

Discussion of the Results.

These results agree fairly well with those obtained by the chloride and bromide methods. The second experiment is more trustworthy than the first. In it, we started with pure metal and the manipulations were so simple that no serious error could could have been made in them. Hence it will only be necessary to consider the end-product i.e the cadmium sulphate. The titration showed that the sulphate was not basic owing to loss of sulphur trioxide, and after deducting the weight of the excess of sulphuric acid

we must have left a weight
of cadmium sulphate which
is equivalent to the metal employed. The question now is,
did it contain anything else and
what would be its effect?
Clearly the effect of water or
any other impurity would be
to lower the atomic weight
found, hence the atomic weight
must be at least as high as
the experiment indicates. As
the cadmium sulphate is deposited, at least the later part of
it is from a strong sulphuric
acid solution, it probably does
not contain any water, and
in this case would fix a maxi-
mum value as well as the

minimum value, and thus
determine the atomic weight.
It might be objected to the
second experiment that the
sulphuric acid found may
have been present as SO_3 and
not as H_2SO_4 as was assumed.
This seems highly improbable,
and even if it were so. the
error introduced would be only
about .03 of a unit in the atomic
weight. As the first determination
was found practically neutral,
it does not apply to it at all.
The most probable conclusion
from these experiments is that
the atomic weight of cadmium
is about 112.35. A more thorough study
of this method would have been made if time
had permitted it

.M

The Oxide Method.

As the chloride and bromide methods and the synthesis of cadmium sulphate all lead to approximately the same high result, it seemed probable that the oxide method which had given a much lower result (Morse & Jones 112.07) must be affected by some error. Accordingly it was examined in the manner about to be described. A set of crucibles was prepared as described by Morse and Jones in their work on this method, and in the present paper under the oxalate method. After they had been heated in a nichel crucibl over

a blast-lamp and weighed,
a weighed piece of cadmium
was introduced into the smaller
inside crucible, and dissolved
in nitric acid with the aid of
heat. An equal quantity of nitric
acid was added to the tare.
The acid was then evaporated
off, and the resulting nitrate
converted into oxide exactly
as has already been described
under the oxalate. The first
experiment was made in this
way and the second one
exactly like it, only the porce
lain crucible used was the
one which had been employed
in the first determination.
The glaze had been removed by the

cadmium oxide of the first de
termination, and before using
for the second one the crucible
was boiled out with nitric
acid, and heated to constant
weight over a blast lamp as before.
Determinations III, IV and V
were made in the same
way except that the small
inner crucible was platinum
instead of porcelain. All weighings were reduced to the vacuum
standard on the basis of 8.34 for
the Sp. Gr. of Cadmium and 8.15 for
the Sp. Gr of cadmium oxide and
8.4 for the brass and 21. for the
platinum weights.
The results are as follows:

	Cd	CdO	At. Wt. Cd.
I.	1.26142	1.44144	112.11
II.	.99785	1.14035	112.04
		Average	112.08

	Cd	CdO	At. Wt. Cd.
III.	1.11321	1.27247	111.84
IV.	1.02412	1.17054	111.91
V.	2.80966	3.21152	111.87
		Average	111.87

The oxides resulting from
these determinations were always
tested for oxides of nitrogen,
sometimes by using meta phenylene
diamine and at other times by
sulphanilic acid and napthylamine
sulphate but no traces were
ever found. The average of
the determinations made in

porcelain crucibles is 112.08
Morse and Jones obtained the
same figure or, if their results
are reduced to the vacuum
standard, 112.11, by the same method
under the same conditions. The
results of the determinations
made in platinum crucibles
are equally constant, but their
average is 111.88 being .20 of a unit
lower. Therefore more oxide is ob
tained when platinum crucibles
are used instead of porcelain ones.
In two cases the platinum crucibles
were were weighed at the end of
the determinations after the
cadmium oxide had been re
moved. Their weight remained
unchanged. The most probable ex

planation of these facts seems to be
that something is retained in the
oxide in both cases, but that the
amount is greater in the determin
ations made in platinum crucibles
than in those in which porcelain
ones were employed. We would
expect this, because in porcelain
crucibles some of the oxide is
absorbed forming a silicate, and
any volatile impurity must be
expelled from this part of the
oxide. Not finding oxides of nitrogen it was thought that gases-
probably nitrogen and oxygen
might be occluded although
Richards and Rogers (Amer. Chem. Jur. 15, 567)
had examined cadmium oxide
prepared from the nitrate and found

only a trace of gas. Accordingly two specimens of cadmium oxide obtained in the above determinations were powdered in an agate mortar and boiled with water for some time in order to remove any adhereing air. They were then dissolved in dilute hydrochloric acid from which the air had been removed by boiling. A small amount of gas was found in each case but not nearly enough to account for the difference of .31 unit in the atomic weight of cadmium between 112.38 and the oxide method. In fact not more than about one sixth of the amount required was found. It may be that the pow

dering of the oxide and then boiling up in water may have been too severe a treatment, and that the greater part of the occluded gas escaped during these processes. It seems that there is at least some error due to occluded gases in methods involving the decomposition of cadmium nitrate to oxide, but no satisfactory idea of its magnitude could be obtained from these two experiments as carried out.

The following experiments were then made and they seem to give definite evidence not only of the existence of an error but also of its magnitude. Carbonate of cadmium was

made by dissolving pure cadmium in nitric acid, adding an excess of ammonia and a small quantity of ammonium carbonate. After standing for some time the cadmium carbonate was filtered filtered off and rejected. The filtrate was treated with an excess of ammonium carbonate and the precipitated cadmium carbonate allowed to digest in it for some time. After washing by decantation several times of the carbonate, was transfered to a funnel containing a porcelain filter-plate, covered with a piece of ashless filter paper of slightly larger diameter, and washed thoroughly.

with water. It was then transferred to a platinum dish, care being taken to avoid contamination with filter paper and heated gently to convert it into oxide. The resulting oxide was powdered in an agate mortar, returned to the platinum dish and heated to incipient whiteness for seven hours in a muffle furnace. The temperature must not be too high, otherwise the oxide will distill readily leaving no residue. The oxide is slightly volatile at good red heat as was observed in trying to make a determination at this temperature by the oxide method.

104

A weighed portion of the oxide which had been prepared from the carbonate in the manner described was dissolved in a weighed porcelain crucible and the resulting nitrate converted into the oxide again by heat just as in the oxide method. This constitutes experiment I. Experiments two and three were made in exactly the same way except that a platinum crucible was used instead of a porcelain one. The results are:

	Initial Wt.	Final Wt.	Gain	Corresponding Grain Alt. Wt.
I.	2.95469	2.95650	.00081	-.24
II.	2.67117	2.67835	.00117	-.34
III.	3.00295	3.00422	.00127	-.38

105

As started with cadmium ox
ide at the beginning of the
experiment, and after passing
to the nitrate, converted it back
into the oxide the weight
should remain unchanged,
if the method is correct.
However, this is not the case,
but a large increase in weight
takes place. The increase
is larger in a platinum crucible
than in a porcelain one, which
accounts the fact that a lower
value of the atomic weight is
found in the oxide method when
they are used. The use of a por
celain crucible therefore elimin
ishes the error, but does not elimin
ate it. The explanation of this has

has already been given. The
oxides obtained in these
three experiments were tested
for occluded gases in the
manner already described, but
only small amounts were
found. Both of those made in
platinum crucibles were tested
for nitrate of cadmium with
brucine and sulphuric acid
with negative results. To show
that the impurity was not converted into an ammonium
salt when the oxide was dissolved in hydrochloric acid, a
slight excess of caustic potash was
added to the solution, the precipitate allowed to subside and the
clear, supernatant liquid tested for

ammonia with hessler's reagent
no ammonia was found.
In order to make these experiments as severe a test as possible, a somewhat higher temperature was employed than had
been done in the five experiments
described under the oxide
method. This was accomplished
by boring out the stopcocks of
the blast-lamp so that a larger
supply of gas was furnished.
The two oxides in the platinum
crucibles seemed to be constant
in weight, but that in the
porcelain crucible seemed to
lose in weight slowly. The
weight given was taken after
four hours blasting, which is

longer and at a higher temperature than was used in any of the five determinations made by the oxide method. If the cadmium oxide prepared from the carbonate retained any carbon dioxide, it would lose weight in being dissolved and reconverted into oxide. The above experiments therefore seem to furnish very strong evidence that there is an error of at least $-.24$ unit in the oxide method when porcelain crucibles are used and $-.39$ of a unit when platinum ones are employed.

If $.24$ of a unit is added to 112.07 the result obtained when porcelain crucibles were used we get 112.31

and adding .39 to 111.87 gives
112.26. Considering the small
number of experiments made,
the fact that they were made
in such a way as to give a low
value (numerically) for the error
rather than a high one, and
also that the error is probably
variable to some extent, especially
when porcelain crucibles are
used, the corrected results agree
as closely with 112.38, the average
of the chloride, bromide and
sulphate (synthetical) methods as
could be expected. It must also
be born in mind that 112.38 is
only to be regarded as an approximation to the atomic weight
of cadmium. The increase in

weight observed in converting
the nitrate back into oxide might
also be explained by assuming
that the cadmium oxide used
in the beginning of the experiments was richer in metal
than the formula CdO indicates
and that the increase in weight
is due to this excess of metal being
changed to oxide. The method
of preparation of the oxide from
the carbonate and the known
properties of cadmium oxide ren
der this view highly improbable,
and the following two bserv
ations render it untenable :
1st. If this were the cause of
the increase, the amount of in
crease would necessarily be the

the same in both platinum
and porcelain crucibles which
is not the case.

2nd. Three grammes of cadmium
oxide made from the carbonate
were dissolved in dilute hydrochloric acid from which the air
had been expelled by boiling.
The oxide which is very com
pact was placed in a glass
bulb which had been blown
at the end of a tube. After
displacing the air by filling
the entire apparatus with
recently boiled water, the exit
of the tube was placed under boiling di
lute hydrochloric acid, and the bulb heated
until the water boiled. It was then turned
over so that the steam dis—

placed nearly all the water.
On removing the flame the
dilute hydrochloric acid at
once filled the bulb. The exit
tube was then quickly placed
under a narrow tube filled with
mercury and inverted over
mercury in a dish. The bulb
was then heated until the
oxide had dissolved. By this
method the gas
would be boiled out
of the solution and
collected in the top
of the narrow tube

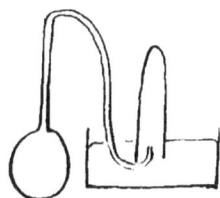

As only a very small amount
of steam and dilute hydrochloric
acid go over at the same time
there is no danger of the gas

formed being absorbed to any considerable extent. It is well to put the oxide into the bulb before the tube is bent. If the hydrochloric acid is too strong, it must be cooled before entering the bulb as otherwise the reaction is too violent and the experiment may be lost. This experiment shows that there is no excess of cadmium present in the oxide employed for no gas was found. If three grammes of the oxide contained enough metal to take up .00126 grms of oxygen, .00016 grms of hydrogen should have been set free, and its volume under ordinary conditions of temperature and pressure would have been about 1.9

cubic centimetres. This experiment
would also have shown the
presence of carbon dioxide if
any had been present.

Discussion of the Oxalate Method.

After having done the
work which has just been described, we are in a position
to turn to the opalate method,
which is the first method
described in this paper. It
involves the decomposition of
cadmium nitrate, and is therefore affected by an error from
this source, only it is not as
large as in case of the oxide
method. If 2.95650 grammes of-

cadmium oxide prepared in a porcelain crucible contain .00081 grammes of impurity an error of –.24 of a unit would be intro duced in the atomic weight as determined by the oxide method or +.10 in case the oxalate method were employed. That is the oxalate should give about 112.48 for the atomic weight of cadmium, but it really gives a very much lower result. Morse and Jones obtained 112.07 ± 035 by it, while Partridge obtained 111.81 ± .035 by it. If we take 112.38 for the atomic weight of cadmium, there appears to be a second error of .44 of a unit in the method as used by Morse and Jones, while

Partridge's result indicates an error of .57 of a unit. Partridge only moistened the oxide obtained from the oxalate with a few drops of nitric acid before making the final heating, and it seems probable therefore that he made no appreciable error on account of the final oxide retaining products of decomposition from cadmium nitrate. The most probable cause of this large error seems probably to be incomplete dehydration of the oxalate, or reduction to metal during the decomposition of the oxalate, and subsequent volatilization of some of it, or a combination of both of

these. The nine determinations given in the earlier part of this paper of course vary so much that they are of no value whatever in determining the atomic weight. The reason that the first four are low is probably due in part to sublimation of cadmium, for, on dissolving the resulting oxide in nitric acid a considerable quantity of metal was noticed in each case. In the others the temperature was kept lower, and the decomposition took a longer time. No metal was observed on taking up in nitric acid. To be certain of what the cause of the error is would

require some very carefully
conducted experiments, but
as there are a number of
much more reliable methods
for determining the atomic
weight of cadmium, it does
not seem desirable to spend
the time required in making
them. It should be mentioned
that Lenssen, in 1866, first employ
ed this method. He made three
determinations. 1.5677 grms of cad
mium oxalate giving 1.0047
grammes of oxide, which gives
a value of 112.043 for the atomic weight
of cadmium oxalate. The difference
between the highest and lowest
determination was .341 of a unit.

Other Methods

A great deal of time was spent in trying to effect a partial synthesis of cadmium bromide in exactly the same manner as had been used in case of cadmium sulphate. No results were obtained because cadmium bromide is slowly volatile at 130°C, the temperature used, retained some hydro-bromic acid even after more than 100 hours of drying. Some work was done on trying to establish the ratio between silver and cadmium by dropping a weighed piece of cadmium into a solution of

silver sulphate, the reaction being:

$$Cd + Ag_2SO_4 = CdSO_4 + 2Ag$$

Silver nitrate cannot be used because it becomes reduced to nitrite even at a temperature of 0°C., as was shown by its reducing action on potassium permanganate, and by the re-action with meta-diamido benzene after the reaction had been completed. The main difficulty with the method is that air must be excluded in order to prevent oxidation and solution of some of the precipitated silver. The silver is perfectly free from cadmium if an excess of silver sulphate is used and the precipitated metal digested

with it for some time.
Since this part of the work
was done, a paper by Mylius
and Fromm (Ber. 1894, 630) appeared
in which one of the reactions
studied was that of cadmium
on silver sulphate. They also
found the resulting silver
free from cadmium. The
method seems very promis-
ing but had to be discontinu-
ed for lack of time.

Conclusion.

I. The work on the oxalate and sulphide methods described in this paper is of no value for determining the atomic weight of cadmium. It does not even enable us to fix an approximate value.

II. There are a number of errors in the chloride and bromide methods as they were used in this work, but they are not very large and partially compensate each other. Their results, 112.383 and 112.396 respectively, may be regarded as approximations to the true value.

III. The synthesis of cadmium sulphate as carried out is of especial value

in fixing a _minimum_ value
for the atomic weight of cadmium. The result is 112.35,
agreeing closely with the
bromide and chloride methods.
IV. There is an error in the oxide
method due to products of decomposition of the nitrate being retained.
Direct experiments gave .39 of a
unit for this, when platinum crucibles were used and .24 of a unit
when porcelain ones were used.
The calculated errors for porcelain
and platinum crucibles are
.30 and .51 of a unit, respectively, if
112.38 is assumed as the atomic
weight of cadmium.
V. The average of the chloride,
bromide, and sulphate methods

is 112.38. This result is to be regarded as _tentative_ and not as final, since the main object of this work has been to find the cause of the discrepancy in some methods employed in determining this constant, rather than to make an atomic weight determination.

Biographical Sketch.

John Emery Bucher was born near Hanover, Pa., August 17, 1872. He entered Lehigh University in 1888 and graduated in 1891. During the past three years he has been a graduate student in the Johns Hopkins University.

Subjects: Chemistry, Mineralogy, and Mathematics.

www.ingramcontent.com/pod-product-compliance
Lightning Source LLC
Chambersburg PA
CBHW021521210326
41599CB00012B/1340